Flo!

HUMOR OUTFITTERS

HUMOR OUTFITTERS

Saratoga Publishing Group Inc.
PO Box 959
Saratoga, Wyoming 82331
(800) 722-6932

ISBN 1-879894-07-6

 Published and distributed exclusively throughout the world by Humor Outfitters, an imprint of The Saratoga Publishing Group Inc. of Saratoga, WY.

Printed in the United States of America

Dear Readers,

For letting me peek into your lives and record the goings-on over the past 25 years, I dedicate this book to you.

I hope this book makes you happy.

Thank you,

Jerry Palen

"We're blaming it on the neighbor's bull."

“OK! Spin it again!”

"I do not! I only drink two cups a meal!"

AG GAZZETTE
HOW TO GET
MORE HELP
FROM YOUR
WIFE
Jerry Daley
10-8

"Tell the dog to stay!"

"Trout food, deer feed, monkey chow,...
Nope, sorry, we don't have any cow feed."

"Hey! I'm serious!"

"It's the first ice of the season and I'm gonna do it right and get it over with!"

"I don't like the way this year is starting out!"

"You know what? We forgot the hay!"

"The dog chased the heifers through the fence again?"

THANK YOU FOR NOT YELLING.

"Oh, stop whining and tell me where reverse is."

"I can't read Elmo's writing. Does that say 'new wrench' or 'new dress'?"

"I'm sorry. I haven't laughed this hard in years!"

"Maybe you ought to start hanging up your clothes."

"Somebody's got to start complaining about these potholes."

"It's cheaper than pesticide and look at the little thugs run!"

"Wait a minute. This isn't dairy supplement. It's chili powder!"

"It sounds like a tiny little combine."

"Of course, I'll always love you...
WAIT a minute!! Where's the pickup?!"

"Flo, what's this charge for bungee cord?"

"I'd say it looks like a tough winter."

"All the smart money is getting rich in the bull market and I own steers!"

"I told him he better get over his cold or I was going to hire someone to help me."

"Your sermon was perfect! Why, you did everything but mention Elmo's name!"

PLEASE
CLOSE
GATE!

"Flo, I'm gonna have to dig us out. You need to go back and get the shovel."

"Where have you been?"

"Your wife sure is good-natured about your yelling all the time."

"He looks a lot bigger in the sale catalog."

"He's got my vote."

"Flo you're doing such a good job, I've decided you should move up the ladder."

"Now, … Who was I yelling at?"

"It broke 10 miles back, but thank goodness, I had plenty of duct tape and bailing wire!"

"Remember?! Remember?! I told you those brakes feel 'sqoooshy'!"

"...out to the heifer pasture. Why?"

"It's been a good rain but he won't be happy until he gets something stuck in it."

"All the baldies go except for the big fat one."

"It'll save you from having to get in and get out opening all those gates."

"Hold that calf right there, Flo! I'll get the medicine and a camera."

"They said 'buy me', 'buy me', and I couldn't turn them down!"

"Quick Flo, bring some bailing wire! All my valuable stuff is trying to escape!"

"Whatever you gave her, give Elmo some."

"I'm a Capricorn and Elmo's a crab."

"I just got 'em! Airbags for the corral!"

"Still trying to adjust the cattle market, Elmo?"

"Flo wheeled him out here. Something about tired of trying to wake him up."

"Cruise?! This is not what I had in mind when I said 'let's take a cruise'."

"Here it is! 'Man falls off haystack yelling at wife!' "

"DUCK!"

"Well, Lassie would go for help!"

"And what am I bid for this part? Guaranteed to fit most anything and yet confound your husband for hours!"

"I know you're not listening! If you were listening, you'd be mad!"

"Never mind! Never mind! I'll open it!"

"OK Flo, one more time and I promise not to yell... Flo?... Flo?"

"Let's get out of here, Flo! He WANTS to loan us money!"

"It just feels like a 'take the ol' wife out for dinner' night!"

"As I get older, my wants are simpler: a pickup with brakes and a gate I can open."

"Do $5,000 bulls always sleep with their feet in the air?"

Order more Jerry Palen books today!

It's Your Gate, Flo!
The further adventures of that wacky farm couple.
ISBN 1-879894-07-6 4 1/4x7" paperback **$5.95**

Flo, Don't Write a Book About It!
Flo and Elmo strike again!
ISBN 0-943255-30-9 4 1/4x7" paperback **$4.95**

STAMPEDE, The First Really Big Book
The BIG (8 1/2x 11") collection of STAMPEDE humor!
ISBN 0-941803-00-7 8 1/2x11" paperback **$6.95**

FLO'S COOKBOOK & CARTOON SERIES

What's Cooking, Flo?
Delicious and easy-to-fix recipes and cartoons!
ISBN 0-943255-31-7 8 1/2x 5 1/2" paperback **$6.95**

What's This Green Stuff, Flo?
Fast & flavorful fruit favorites & very veggie recipes!
ISBN 1-879894-03-3 8 1/2x 5 1/2" paperback **$6.95**

You're Hot Stuff, Flo!
Simple, savory, and spicy Southwestern specials!
ISBN 1-879894-08-4 8 1/2x 5 1/2" paperback **$7.95**

Other books illustrated by Jerry Palen

A Wide Spot in The Road by Bob Budd
Humorous tales of growing up in a small town.
ISBN 0-943255-29-5 6x9" paperback **$9.95**

Send Fresh Horses by Bob Budd
Hilarious episodes in the life of a modern-day cowboy.
ISBN 0-943255-01-5 6x9" paperback **$8.95**

The Beef Lover's Guide by Chriss McNaught
Great BEEF recipes for good taste and good health!
ISBN 0-943255-27-9 6x9"spiral-bound **$12.95**

To order: call Humor Outfitters at
(800) 722-6932